Fantastic ✦ Physical ✦ Science ✦ Experiments

Energizing Science Projects with Electricity and Magnetism

Robert Gardner

Enslow Elementary

an imprint of

 Enslow Publishers, Inc.

40 Industrial Road   PO Box 38
Box 398   Aldershot
Berkeley Heights, NJ 07922 Hants GU12 6BP
USA   UK

http://www.enslow.com

Enslow Elementary, an imprint of Enslow Publishers, Inc.

Enslow Elementary® is a registered trademark of Enslow Publishers, Inc.

Library of Congress Cataloging-in-Publication Data

Gardner, Robert, 1929–
    Energizing science projects with electricity and magnetism / Robert
Gardner.—1st ed.
        p. cm. — (Fantastic physical science experiments)
    Includes index.
    ISBN-10: 0-7660-2584-5 (hardcover)
    1. Electricity—Experiments—Juvenile literature. 2. Magnetism—
Experiments—Juvenile literature. I. Title. II. Series
    QC527.2.G385 2006
    537'.078—dc22   2005018730

ISBN-13: 978-0-7660-2584-4

Printed in the United States of America

10 9 8 7 6 5 4 3 2

To Our Readers: We have done our best to make sure all Internet Addresses
in this book were active and appropriate when we went to press. However, the
author and the publisher have no control over and assume no liability for the
material available on those Internet sites or on other Web sites they may link to.
Any comments or suggestions can be sent by e-mail to comments@enslow.com
or to the address on the back cover.

Illustration credits: Tom LaBaff

Cover illustration: Tom LaBaff

Contents

(Experiments with a 🎀 symbol feature **Ideas for Your Science Fair.**)

Introduction

More than 250 years ago, Benjamin Franklin experimented with electricity. Only about 100 years ago did electricity become widely used to light American homes and streets. Today, people around the world depend on electricity in their daily lives. In how many ways do you use electricity?

Chinese sailors used magnets as compasses 2,000 years ago. But it was 1820 before Hans Christian Oersted discovered that electricity and magnetism are related. You will repeat his discovery later in this book. You will also use electricity to make a magnet. And you will do much more as you uncover the secrets of magnets and electricity.

Entering a Science Fair

Most of the experiments in this book (those marked with a 🎗 symbol) are followed by ideas for science fair projects. Judges at science fairs like experiments that are creative. So do not simply copy an experiment from this book. Expand on one of the ideas suggested, or think of a project of your own. Choose a topic you really like and want to know more about.

Then your project will be more interesting to you. Your curiosity can lead to a creative experiment that you plan and carry out.

Before entering a science fair, read one or more of the books listed under Further Reading. They will give you helpful hints and lots of useful information about science fairs.

Safety First

To do experiments safely, always follow these rules:

① Do all experiments under adult supervision.

② Read all instructions carefully. If you have questions, check with the adult.

③ Be serious when experimenting. Fooling around can be dangerous to you and to others.

④ Keep the area where you work clean and organized. When you have finished, clean up and put all the materials away.

⑤ Never experiment with electric wall outlets.

1. Electric Charges

Electricity is made of two kinds of charged particles. To separate the two kinds of charges, you can do an experiment. This experiment should be done on a dry cool or cold day. It might also be done in a well air-conditioned room. Otherwise, skip to Experiment 4.

Let's Begin

❶ Use a string and tape to hang a plastic ruler from a cabinet or table.

❷ Rub the ruler all over with a paper towel. This will put electric charges on the ruler. Charge an identical ruler in the same way.

❸ Slowly bring the second ruler near the hanging ruler. What happens?

❹ Hold the unfolded paper towel near one side of the hanging ruler. What happens?

❺ Charge the hanging ruler again by rubbing it with the paper towel. Rub a tall drinking glass with the paper towel. Hold the glass near the ruler. What happens?

at Rest

Things you will need:
- ✔ string
- ✔ tape
- ✔ 2 plastic rulers
- ✔ cabinet or table
- ✔ paper towel
- ✔ tall drinking glass
- ✔ balloon

6 Charge the hanging ruler again. Blow up a balloon. Rub it all over with the paper towel. Hold it near the ruler. What happens?

Try to explain all that you have seen.

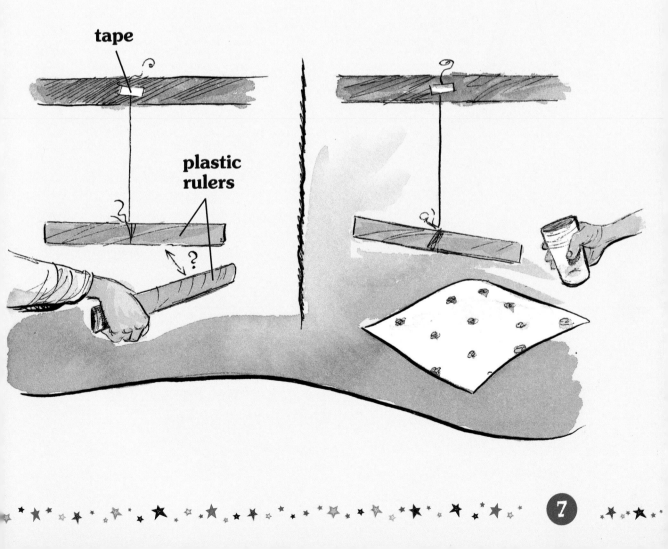

tape

plastic rulers

Electric Charges at

You charged the two rulers the same way: You rubbed them with a paper towel. This put the same charge on each ruler.

What happened when you held one ruler close to the other ruler? The rulers pushed away from each other. This shows that like charges push away from each other.

The opposite happened with the paper towel. It attracted the ruler. Why did this happen?

You already know there was one type of charge on the ruler. Since the paper towel *attracted* the ruler, it must have had an opposite charge. Opposite charges attract each other.

The glass rubbed with the towel attracted the ruler, so the ruler and glass had opposite charges. Depending on its makeup, the balloon may have attracted or repelled (pushed away from) the ruler. If it repelled the ruler, the balloon had the same kind of charge as the ruler. If the two were attracted, the balloon and ruler had opposite charges.

The electric charges you separated in this experiment are called static charges. They do not

Rest: An Explanation

leave the material they are on. Later, you will find that electric charges can flow along wires.

Rubbing the ruler with a paper towel separated the two kinds of charge (x and y).

Idea for Your Science Fair

★ Can you put electric charges on plastic rulers, balloons, and other things by rubbing them with woolen and silk cloths? If you can, are the charges the same or different than the ones made with the paper towel? Why can't you charge metal objects this way?

2. Naming the Two

There are two kinds of charge. But how can you tell one kind from the other? Benjamin Franklin named them. The charge on glass rubbed with silk he called positive (+). The charge on rubber rubbed with wool or fur he called negative (−). Do this experiment on a dry, cool or cold day. It might also be done in a well air-conditioned room. Otherwise, skip to Experiment 4.

Let's Begin!

Things you will need:
- ✔ plastic ruler
- ✔ thread or string
- ✔ tape
- ✔ cabinet or table
- ✔ paper towel
- ✔ drinking glass
- ✔ silk cloth

❶ Hang a plastic ruler from a thread or string as you did in Experiment 1. Rub the ruler with a paper towel.

❷ Rub a drinking glass with a piece of silk. Slowly bring the glass near the charged ruler. What happens? Is the ruler attracted or repelled by the glass?

❸ You know the charge on the glass is positive. What is the charge (+ or −) on the ruler?

Kinds of Charge

4 Predict the charge on a glass rubbed with a paper towel. Predict the charge on a balloon rubbed with a paper towel.

5 Test your predictions by doing experiments. Were your predictions right?

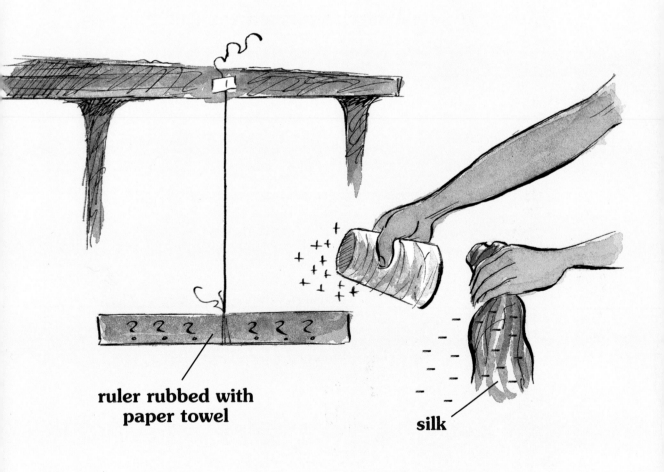

ruler rubbed with
paper towel

silk

Naming the Two Kinds of

You know that like charges repel one another. And you know that opposite charges attract. The glass rubbed with silk had a positive charge. (Remember: Franklin decided that the charge on glass rubbed with silk would be called positive.) The silk became negatively charged. You probably found that the glass attracted the charged ruler. This tells you that the ruler had a negative charge (after it had been rubbed with a paper towel).

The glass rubbed with paper probably attracted the plastic ruler. If it did, its charge was positive, just like the glass rubbed with silk.

If the balloon attracted the negatively charged ruler, the balloon carried a positive charge. If it repelled the ruler, it was negatively charged.

Your results may have been different. If so, they may have been affected by something you will discover in the next experiment.

Charge: An Explanation

Opposite charges attract!

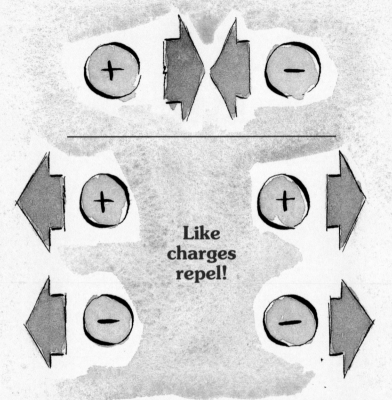

Like charges repel!

Ideas for Your Science Fair

★ What is an electroscope? Find out and then build one. Use it to determine the charge (+ or −) on various objects rubbed with silk, a paper towel, or wool.

★ Why are cool, dry days best for charging objects?

3. Charge Near an

Do this experiment on a cool or cold, dry day. Or do it in a well air-conditioned room. Otherwise, skip to Experiment 4.

Let's Begin!

❶ Tear some paper into tiny pieces. Let them fall onto a table or counter. Put negative charge on a plastic ruler by rubbing it with a paper towel.

Things you will need:
- ✔ paper
- ✔ table or counter
- ✔ plastic ruler
- ✔ paper towel
- ✔ Rice Krispies cereal
- ✔ aluminum foil
- ✔ thread
- ✔ scissors
- ✔ tape
- ✔ cabinet or table

❷ Bring the charged ruler near the pieces of paper. What happens? Can you explain why it happens?

❸ Repeat the experiment using Rice Krispies cereal. Are the results the same?

❹ Repeat the experiment using tiny pieces of aluminum foil rather than paper. Are the results the same?

Uncharged Object

❺ Cut a piece of thread about one foot long. Press a tiny piece of aluminum foil firmly around the end of the thread. Tape the other end of the thread to a cabinet or table. The aluminum foil should be free to move.

❻ Charge a plastic ruler by rubbing it with a paper towel. Slowly bring the ruler near the hanging aluminum foil. Watch carefully. What happens? Try to explain what you see.

pieces of paper

aluminum foil

Charge Near an Uncharged

All matter has electric charges. There are charges in the small pieces of paper, the cereal, and the aluminum. But there are as many positive charges as there are negative charges. The + and − charges are equal in number. We say the pieces are neutral. They have no excess positive or negative charge.

You put the negatively charged ruler near the neutral pieces of paper, the cereal, and the aluminum. Positive charges in the pieces were attracted to the ruler. Negative charges were repelled. The negative charges moved to the side of the pieces

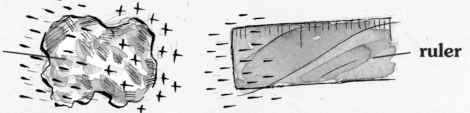

paper or aluminum

ruler

farthest from the ruler. This left the side closest to the ruler positively charged. The pieces were attracted to the ruler because positive charges were closer to the ruler than the negative charges. The closer the charges, the bigger the attraction or repulsion. The farther apart the charges, the smaller the attraction or repulsion.

Object: An Explanation

The neutral hanging piece of aluminum was attracted to the ruler. But then the aluminum was suddenly repelled by the ruler. Why? When the ruler touched the aluminum, it shared some of its negative charges with the aluminum. After sharing charges, both ruler and aluminum had excess negative charge. As a result, they repelled one another.

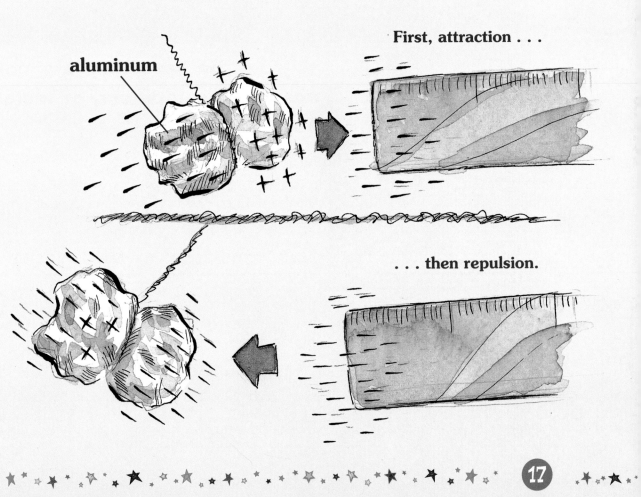

aluminum

First, attraction . . .

. . . then repulsion.

4. Magnets and a

Magnets have two poles—north and south. How can you tell which pole is which? Do an experiment to find out.

Let's Begin!

Things you will need:
- ✔ thread
- ✔ tape
- ✔ furniture
- ✔ 4 or 5 ceramic magnets (obtain from an electronics or hardware store)
- ✔ magnetic compass
- ✔ marking pen

❶ Using thread and tape, hang several ceramic magnets in a room. If you have a bar magnet, hang it too. Hang them from shelves, tables, chairs, or whatever is practical. **Be sure they are not near TVs, computers, or metal objects.** Let them come to rest. Are they all turned the same way? Do they all face the same direction?

❷ Hold a magnetic compass. Keep it several feet from any magnet or metal object. Do the magnets face the same way that the compass needle points?

A compass needle is a small bar magnet. It points in a northerly direction. The end pointing north is the magnet's north pole. The other end is its south pole.

The side of a hanging magnet facing north is its north pole. The other side is its south pole.

Compass

❸ Carefully hold one of the hanging magnets in one hand. Be sure it has not turned. Use a marking pen to write N on the side that faces north. Do this for each hanging magnet. Save the magnets for the next experiment.

bar magnet

compass

magnet

north
pole

Magnets and a Compass:

If free to turn, all magnets come to rest along a north-south line. The end or face that is farthest north is called the magnet's north pole. It is also called its north-seeking pole. The other end or face is the magnet's south or south-seeking pole. However, many metal objects are magnetic. They are attracted by magnets. They may make the magnet point in a different direction. That is why the hanging magnets should not be near any metal objects. (Magnets can destroy computers and other electronic devices, so keep magnets far from them.)

A compass needle is simply a small bar magnet free to turn. Its north-seeking pole always points in a northerly direction. However, it rarely points toward Earth's geographic North Pole, which lies almost directly under the North Star. It usually points either slightly east or slightly west of Earth's North Pole. You will see why this is true after the next experiment.

An Explanation

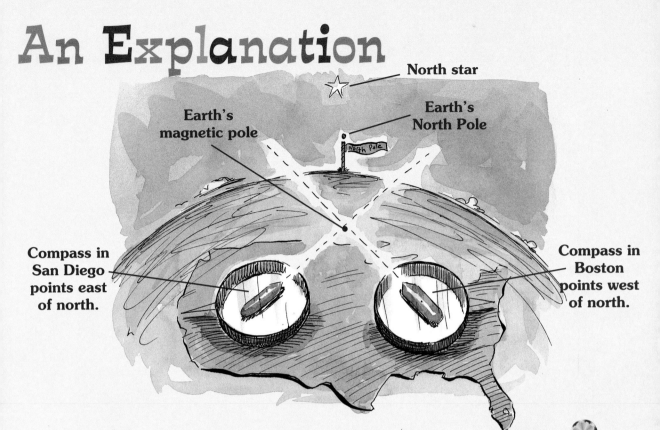

North star

Earth's North Pole

Earth's magnetic pole

North Pole

Compass in San Diego points east of north.

Compass in Boston points west of north.

Ideas for Your Science Fair

★ Take a magnetic compass outside at night. Have an adult help you locate the North Star. Use a flashlight to look at your compass. Does it point in the direction of the North Star? Or does it point east or west of the North Star?

★ Make a compass needle. Stroke a sewing needle with a strong magnet. Always stroke it in the same direction. How can you support the needle so as to make a compass?

5. Magnets Attract

In Experiment 4, you printed an *N* on the north pole of each hanging magnet. What letter should you print on the other face of each magnet?

Let's Begin!

❶ Hold a magnet in each hand. Slowly move the north pole of one magnet toward the south pole of the other magnet. What happens?

❷ Next, slowly move the south pole of one magnet toward the north pole of the other magnet. What happens?

❸ Slowly move the north pole of one magnet toward the north pole of another magnet. What happens?

❹ Slowly move the south pole of one magnet toward the south pole of another magnet. What happens?

❺ From all these experiments, what can you conclude?

❻ Make a prediction: What will happen if you slowly bring the north pole of one of your magnets near the north end of a compass needle? Try it! Were you right?

or Repel

7 Make another prediction: What will happen if you slowly bring the north pole of one of your magnets near the south end of a compass needle? Try it! Were you right?

> Things you will need:
> ✔ 2 ceramic or bar magnets
> ✔ magnetic compass

Magnets Attract or Repel:

You labeled the north-seeking poles of your magnets. The opposite faces would be the south-seeking poles (S).

You moved the north pole of one magnet toward the south pole of another. The magnets attracted each other. You moved the south pole of one magnet toward the north pole of another magnet. They also attracted each other. But two north poles repelled each other. And two south poles repelled each other. As you see, like poles (N and N, or S and S) repel. Unlike poles (N and S) attract.

A compass needle works because Earth acts as if it has a giant magnet inside. The south-seeking pole of Earth's magnet is in northern Canada. That is why the north-seeking pole of compasses points toward a place in northern Canada, not toward Earth's North Pole.

Early scientists discovered just what you have discovered. Like magnetic poles repel, and unlike poles attract. They knew, too, that like electric charges repel, while unlike charges attract. This knowledge made them wonder: Could electricity and magnetism be related?

An Explanation

N–S

Attract

N–N

repel

S–S

repel

Earth's North Pole

equator

South Pole

Ideas for Your Science Fair

★ Build a model to show that Earth's magnetic pole in northern Canada must be a south-seeking pole.

★ Does the force between two magnetic poles change as the distance between them changes? Do an experiment to find out.

6. Magnetic and

Objects with no excess charge (+ or −) may be attracted to a charged object. You discovered this if you did Experiment 3. Can objects that are not magnetic be attracted to a magnet?

Let's Begin!

❶ Gather a number of common objects such as pieces of paper, wood, plastic, various metals (copper, aluminum, brass, nails, coins, paper clips, etc.), magnets of various kinds (bar, horseshoe, ceramic, etc.), as well as other things.

❷ Test each object by bringing a magnet near it. Place the objects you test in one of three piles: (a) magnets—things that are attracted by one pole of a magnet and repelled by the other pole of the magnet; (b) magnetic—things that are attracted but

> **Things you will need:**
> ✔ common objects: pieces of paper, wood, glass, plastic, keys, bolts, pins, needles, stones, various metals (copper, aluminum, brass, nails, coins, paper clips, etc.)
> ✔ magnets of various kinds (bar, horseshoe, flat, round, etc.)
> ✔ ceramic or bar magnet

Nonmagnetic

not repelled by the magnet; (c) nonmagnetic—things that are neither attracted nor repelled by the magnet. What kinds of things do you find in each pile?

❸ Stroke one or more of the long, narrow objects in pile (b) with a magnet. Always stroke in the same direction with just one pole of the magnet. How can you test to see if the object has become a magnet? Were you able to change any of the things in pile (b) into magnets?

items to be tested

Magnetic and Nonmagnetic:

Most of the objects you tested were probably in pile (c). Pile (a) consisted of any magnets you may have had, whatever their shape or size. Pile (b) contained metals such as iron, steel, cobalt, or nickel. These metals are magnetic. Other metals, such as aluminum, are not. Magnetic metals are those attracted by a magnet even though they are not magnets. This is similar to what you saw in Experiment 3. An uncharged object may be attracted to a charged object. Charges in the object are attracted or repelled. Because the attracted charges are pulled closer, the overall effect is attraction.

In the same way, a magnet held near magnetic objects can make them magnetic. Atoms of iron or steel, for instance, are themselves tiny magnets. The magnetic poles of iron atoms are attracted by a magnet. They make the iron object stick to the magnet. But the magnetism (the magnetic attraction) does not last. The iron objects are not magnets after the magnet is removed. They do not have north and south poles.

An Explanation

Normally, the magnetic poles of iron atoms are not lined up. However, if stroked with one pole of a magnet, they can become magnets. The stroking causes the atoms to line up as shown: N-S-N-S-N-S. This strengthens the magnetic effect, turning the magnetic material into a long-lasting magnet.

Stroking a magnetic object can make it into a magnet. The atoms line up, making the object into a magnet.

pin

Detail Pin head

Before

After

Ideas for Your Science Fair

★ Design some games that use magnets.
★ Figure out a way to test the strength of magnets.

7. Battery, Bulb, and Wire

An electric circuit lets charge flow from one end of a battery to the other. The two ends of a battery can be connected to an incandescent bulb by wires. This loop is called a circuit. Such a circuit allows electric charges to move through the bulb, making it glow.

Let's Begin!

❶ Gather two bare copper wires, a D-cell battery, and a flashlight bulb. Ask a friend to help you. An extra pair of hands will be useful.

❷ Can you make the bulb light? In how many different ways can you make the bulb light?

Things you will need:
- ✔ 2 bare copper wires about 6 inches long
- ✔ D-cell battery (1.5 volt)
- ✔ incandescent flashlight bulb rated 1.5–6 volts
- ✔ a friend

❸ Are there special places where the wire must touch the bulb?

❹ Are there special places where the wire must touch the battery?

❺ How many ways can you make the bulb light using just one wire?

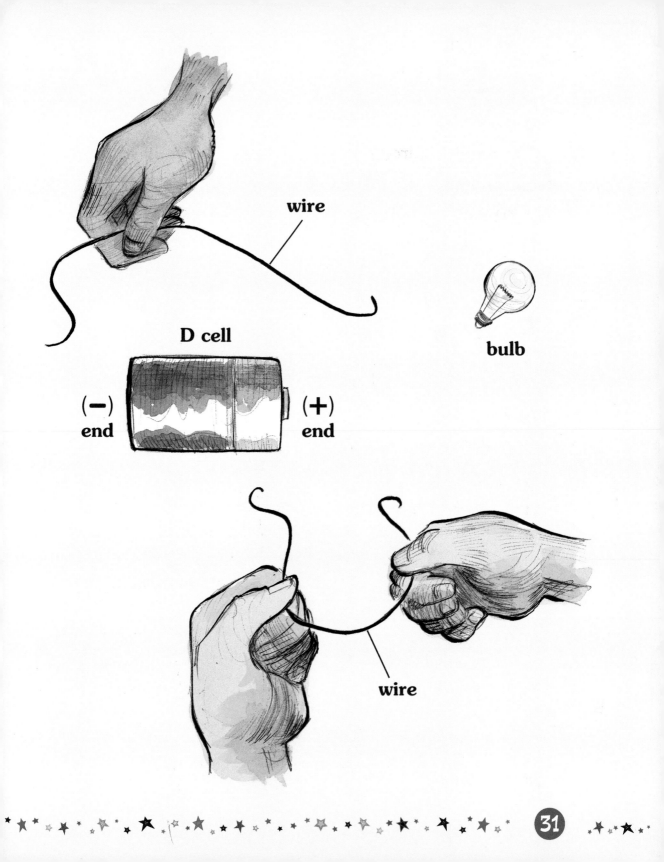

wire

bulb

D cell

(−)
end

(+)
end

wire

Battery, Bulb, and Wire:

Chemicals in a battery cause negative charges to gather at one end. Positive charges gather at the other end. The positive end of the battery can pull negative charges (electrons) along one or more wires. The charges move along the wire from the negative end to the positive end. If the charges also move through a bulb, they can make the bulb glow.

It is easy to light the bulb using two wires. One wire must touch the battery's negative (–) metal end. The second wire must touch the battery's positive (+) metal end. One wire must also touch the small metal tab at the base of the bulb. The second wire must touch the metal side of the bulb (see drawing). The metal tab at the bulb's base and the metal on the bulb's side are not connected.

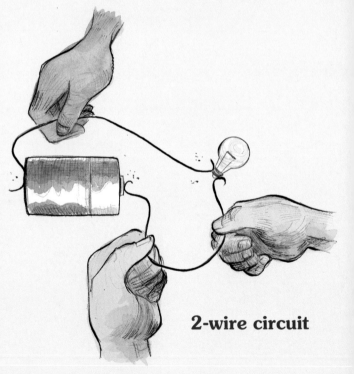

2-wire circuit

An Explanation

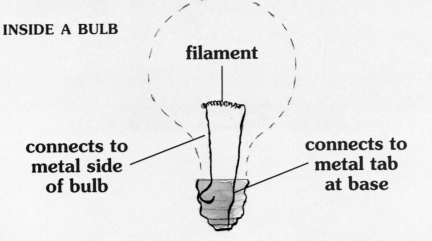

filament

connects to
metal side
of bulb

connects to
metal tab
at base

Inside the bulb is a metal filament. One end of the filament is connected to the metal side of the bulb. The other end is connected to the metal tab at the bulb's base. The charges have to travel through the bulb's filament. When they do, the bulb lights.

You can light the bulb with just one wire. Simply put the base or side of the bulb directly against either end of the battery.

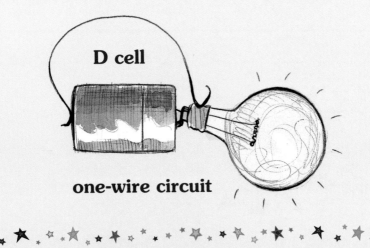

D cell

one-wire circuit

8. Build a Circuit

A switch turns a circuit on and off. You can build a circuit that has a switch (see drawing).

Let's Begin!

❶ Coil one end of each of two wires (#1 and #2). Use a strong, wide rubber band to hold these wire ends firmly against the ends of a D-cell battery.

Things you will need:
- ✔ 3 bare copper wires about 6 inches long
- ✔ D-cell battery
- ✔ 2 strong, wide rubber bands
- ✔ 3 thumbtacks
- ✔ 2 soft pieces of wood or thick wood shingles
- ✔ metal paper clip
- ✔ flashlight bulb rated 1.5–6 volts
- ✔ clothespin

❷ Slide the other end of wire #1 under a thumbtack. Press the thumbtack into a soft piece of wood or a thick wood shingle.

❸ Slide the other end of wire #2 under another thumbtack. This thumbtack should also have a paper clip beneath it. Press that thumbtack firmly into a second piece of wood. The thumbtack will hold both paper clip and wire in place.

❹ Wrap one end of wire #3 around the side of a flashlight bulb. Use a clothespin to hold the wire firmly against the side of the bulb. Use a strong rubber band to hold the clothespin in place. The base of the bulb must press against the thumbtack connected to the battery by wire #1.

5 Slide the other end of wire #3 under a thumbtack. Press this thumbtack into the piece of wood near the free end of the paper clip. How can you make the bulb light?

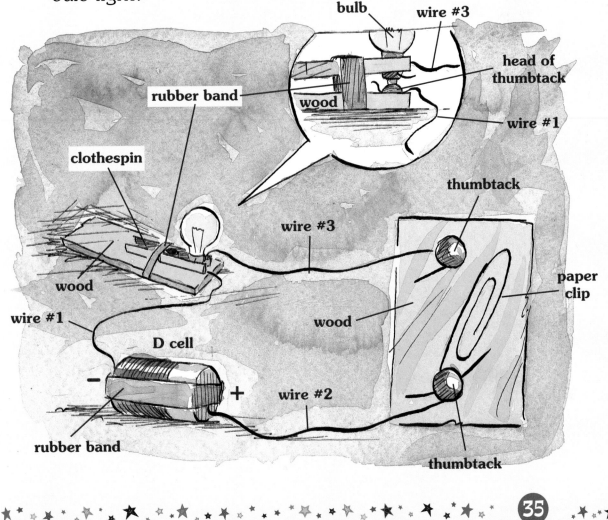

Build a Circuit:

The circuit you built contains a switch. The switch is the paper clip that can connect wire #2 to wire #3. By turning the paper clip so that it touches both thumbtacks, you close the circuit. Closing the circuit allows electric charge to move from the battery through wire #1 to the bulb. From there it can flow to the switch through wire #3. It can move back to the battery through wire #2.

To open the circuit, simply turn the paper clip so that it does not touch both thumbtacks. This creates a gap in the circuit. Charge can no longer flow along wires from one end of the battery to the other. The bulb goes out.

Ideas for Your Science Fair

★ There are probably switches in your home that allow you to turn a light on or off from different places. Build a sample combination switch, using similar materials from this experiment, that will close or open the circuit you built from two different locations.

An Explanation

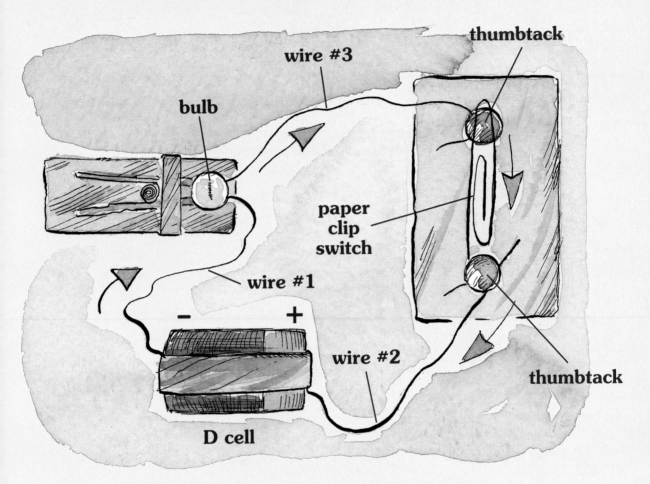

thumbtack

wire #3

bulb

paper
clip
switch

wire #1

thumbtack

- +

wire #2

D cell

★ What is a short circuit? Use a wire to short-circuit the circuit you built.

★ What are parallel circuits? What are series circuits? Make a parallel circuit with two or more bulbs in it. Make a series circuit with two or more bulbs in it.

9. Electricity and Magnetism

In 1820, Hans Christian Oersted discovered a connection between electricity and magnetism. You can make the same discovery.

Let's Begin!

Things you will need:
- ✔ an adult
- ✔ 2–3 feet of insulated copper wire (coated with plastic or enamel)
- ✔ knife or wire stripper or sandpaper
- ✔ a friend
- ✔ magnetic compass
- ✔ D-cell battery

❶ **Have an adult** remove about an inch of insulation from each end of a long piece of insulated copper wire. A knife or wire stripper can be used for plastic-coated wire. Sandpaper can be used to remove the enamel from an enamel-coated wire.

❷ Have a friend hold the middle of a long piece of the insulated wire against the top of a magnetic compass. The wire should line up with the compass needle (see drawing).

❸ Briefly connect the bare ends of the wire to opposite ends of a D-cell battery. Do not connect the circuit for more than a few seconds. Charges will flow along the wire from one end of the battery to the other. What happens to the compass needle? What does this tell you?

4 Place the wire beneath the compass. Briefly touch the wires to the ends of the D-cell battery. What is different this time? What does this indicate?

5 Turn the battery around. Then repeat the experiment. What is different this time? Why do you think it is different?

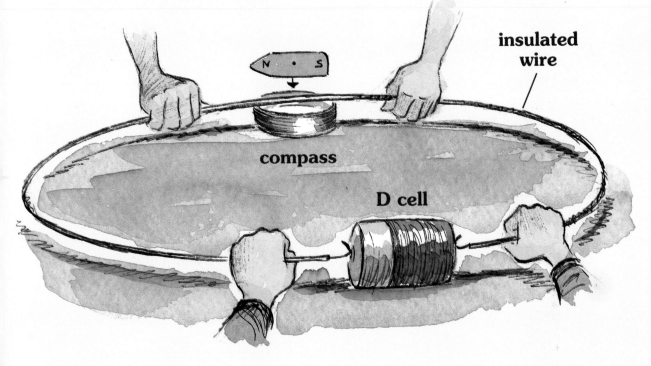

insulated wire

compass

D cell

Electricity and Magnetism:

A flow of electric charge is called an electric current. As you discovered, an electric current in a wire creates a magnetic effect. The magnetism made by the current is much stronger than Earth's magnetism. The wire and compass needle were lined up before the battery was connected. When there was a current in the wire, the compass needle (magnet) turned. It pointed east or west across the wire (perpendicular to it).

Suppose the needle turned west when it was below the wire. It would point east when placed above the wire. This indicates that the magnetism caused by the current circles (goes around) the wire.

Turning the battery around makes current (electric charges) move in the opposite direction. This turns the magnetic effect around. If the current caused the needle to point west before, it will now make it point east.

As you probably found in Experiment 6, a steel nail can be made into a magnet. What do you think you can make by wrapping insulated wire around a steel nail? Find out in Experiment 10.

An Explanation

Ideas for Your Science Fair

★ Do an experiment to show that moving magnets can produce an electric current.

★ Evidence suggests that Earth's magnetic poles have been reversed several times during Earth's existence. Can you explain how this could have happened?

10. An Electromagnet

Remember Oersted's discovery? Soon after, the scientific connection between electricity and magnetism was put to good use. This experiment will show you one use.

Let's Begin!

❶ Obtain a large steel nail. Hold the nail near a compass to be sure it has not been magnetized.

❷ Obtain about 5 feet of insulated copper wire. **Ask an adult** to remove about an inch of insulation from each end of the wire (see Experiment 9).

❸ Leave straight about 8 inches at each end of the wire. Wrap the rest tightly around the nail. Be sure to keep wrapping the wire in the same direction.

❹ Hold one end of the nail. Ask a friend to hold the ends of the wire firmly against opposite ends of a D-cell battery for just a few seconds. While the circuit is closed, touch the other end of the nail to a pile of paper clips. Lift the nail. What happens?

❺ Have your friend open the circuit. What happens?

6 Repeat the experiment to check the results. Do not leave the battery connected for more than a few seconds. The battery will wear out quickly, because a large current flows through the wire.

Things you will need:
- ✔ large steel nail
- ✔ magnetic compass
- ✔ 5 feet of insulated copper wire (coated with plastic or enamel)
- ✔ an adult
- ✔ knife or wire stripper or sandpaper
- ✔ a friend
- ✔ D-cell battery
- ✔ paper clips

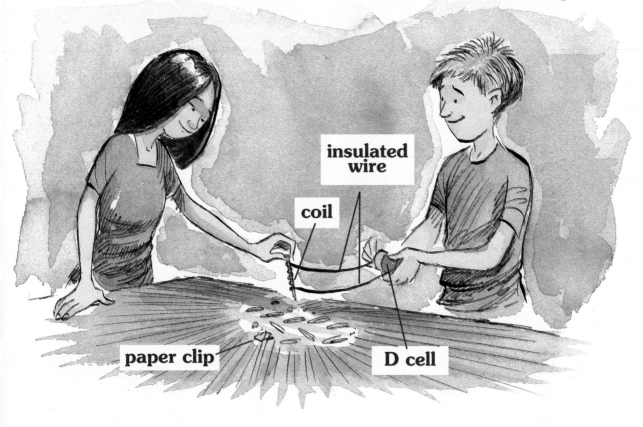

insulated wire

coil

paper clip

D cell

An Electromagnet:

You made an electromagnet. Remember, electric current is charge flowing through a wire. Electric current produces a magnetic effect around the wire. Many turns of wire magnify the effect. There was a large electric current in the coil of wire connected to the battery. The nail became magnetized (became a magnet) when current flowed through the wire coil. Any steel or iron core inside a wire coil carrying an electric current will be magnetized.

The nail lost its magnetism when the circuit was disconnected. The magnetism comes from the electric current. When there is no charge moving through the wire, most of the magnetism disappears.

Huge electromagnets are found in junkyards and other places. They are used to lift and move heavy piles of steel. When the current is switched off, the magnetism disappears. The heavy steel objects fall. Electromagnets are also used to separate magnetic objects from nonmagnetic ones. Aluminum objects are not attracted to the electromagnet. So an electromagnet can be used to separate steel from aluminum.

An Explanation

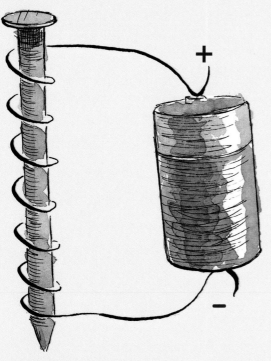

N

+

−

S

Electromagnet

Ideas for Your Science Fair

★ Show that when a wire carrying a current is placed between the poles of a strong horseshoe magnet, there is a force on the wire. Use that force to create a small electric motor.

★ Use insulated wire and a magnetic compass to build a meter that can detect an electric current.

Words to Know

attraction—A force that pulls objects toward one another. Opposite electric charges (+ and −) attract. So do opposite magnetic poles (N and S).

battery (electric cell)—A device in which chemical reactions cause opposite electric charge to gather at separate ends. When the ends are connected by metal, negative charges (electrons) travel from the negative end to the positive end.

electric circuit—A closed path that allows charge to flow from one end of a battery to the other. If the path is broken, perhaps by a switch, the circuit is open, and charge stops flowing.

electricity—Tiny but powerful motion that happens because of electric charges.

electromagnet—A wire coil wound around an iron or steel core. When current flows in the wire, the core becomes a magnet.

magnet—An object that will line up in a north–south direction when free to turn. One end will be attracted to the north, the other end to the south.

magnetic compass—A small bar magnet that is free to turn. Its north-seeking pole will point in a northerly direction.

repulsion—A force that pushes objects away (repels them) from one another. Electric charges that are alike (− and −, or + and +) repel. So do like magnetic poles (N and N, or S and S).

switch—A device used to open and close an electric circuit.

Further Reading

Bartholomew, Alan. *Electric Mischief: Battery-Powered Gadgets Kids Can Build.* Toronto: Kids Can Press, 2002.

Bombaugh, Ruth. *Science Fair Success, Revised and Expanded.* Springfield, N.J.: Enslow Publishers, Inc., 1999.

Farndon, John. *Electricity.* New York: Benchmark Books, 2001.

Good, Keith. *Zap It!: Exciting Electricity Activities.* Minneapolis: Lerner Publications Company, 1999.

Lauw, Darlene, and Lim Cheng Puay. *Electricity.* New York: Crabtree Publishing, 2002.

Nankivell-Aston, Sally, and Dorothy Jackson. *Science Experiments with Electricity.* New York: Franklin Watts, 2000.

Internet Addresses

The Exploratorium. *Science Snacks.* <http://www.exploratorium.edu/snacks>.

Click on "Snacks by Subject."
Then click on "Electricity" or Magnetism."

NASA. *The NASA SCIence Files.* "Electricity Activities." <http://scifiles.larc.nasa.gov/text/kids/D_Lab/acts_electric.html>.

Index